PROGRESS OF STATISTICS;

READ BEFORE THE

AMERICAN

Geographical & Statistical Society,

AT THE

ANNUAL MEETING IN NEW YORK,

DEC. 1, 1859.

By JOS. C. G. KENNEDY, A. M.,

Superintendent of the United States Census; Corresponding Member of the Society, and of the Royal Statistical Commission of Belgium, and London and Dublin Statistical Societies, etc., etc.

NEW YORK:

J. F. TROW, PRINTER, 50 GREENE STREET.

1861.

PROGRESS OF STATISTICS;

READ BEFORE THE

AMERICAN

Geographical & Statistical Society,

AT THE

ANNUAL MEETING IN NEW YORK,

DEC. 1, 1859.

By JOS. C. G. KENNEDY, A. M.,

Superintendent of the United States Census; Corresponding Member of the Society, and of the Royal Statistical Commission of Belgium, and London and Dublin Statistical Societies, etc., etc.

NEW YORK:
J. F. TROW, PRINTER, 50 GREENE STREET.
1861.

VII.—*The Origin and Progress of Statistics.* BY JOSEPH C. G. KENNEDY, COR. MEM. A.G.S.S., Superintendent of the United States Census.

Read December 1, 1859.

The human mind dwells with satisfaction upon ascertained results, and finds pure enjoyment in the contemplation of truths which evince a progressive knowledge respecting the real condition of the human family.

Theories, by exciting the mental faculties, create an interest according to the novelty or magnitude and importance of their topics; but, based as they are on uncertainty, exercise only, but do not satisfy the mind, which finds repose in truths alone; a repose more or less affected by the degree of confidence with which we are enabled to determine what is truth, particularly in its relation to ourselves.

Disturb the mind with doubt or harrow it with uncertainty, man's convictions afford but little real comfort, and however correct they may be in the abstract, they yield but little pleasure, while their weakness deprives him of the ability or dispotion to exert an influence by the impress of his sentiments upon others. Fanaticism may inspire confidence and courage, but its influence is apt to be temporary and limited. With regard to matters of conscience, and the determination of truths respecting religious trust, the same *force of conviction* as would naturally arise from actual demonstration, must exist to inspire faith and confidence, but conclusions in that case rest, in the nature of things, upon what is not susceptible of demonstration by any appeal to facts, and the mind settles upon what it deems most consistent with its convictions of a superior Providence and its relations to humanity, upon tradition or Divine Revelation, calling into play the comforting influences of faith. Religious convictions force themselves upon the belief because of our relations to what we know to be true. Man perceives himself infinitely above all other creatures which inhabit the globe he occupies; he realizes that all created things are constructed for his uses, and discovers his ability to overcome all physical obstacles to the attainment of his wishes, while he has not the power to construct or create the most trifling object in nature; himself the work of an intellect or force as immeasurably above his own as his is superior to those beneath him. He can tunnel mountains which the concentrated energy of the world could not raise; he can make rivers subservient to his uses and span them with an arch, but cannot create a rivulet; he perceives a vast luminary cheering successive generations with undiminished light and heat, and the harmony and uniformity of nature maintained for ages for him alone, realizing at the same time that he is but as a speck upon the earth, the memory of which departs from the knowledge of men centuries before other objects formed centuries in advance, show signs of decay; while the earth itself which we inhabit is but an atom in comparison with what we know of the rest of creation. With such truths patent to the observation of every one of ordinary comprehension, and perceiving illustrations in nature harmonizing with the cheering idea, man deduces the conviction that he is the creature of a higher power who controls his destinies, believes in a future existence, and hopefully settles upon some form of natural or revealed religion, according to surrounding circumstances. How beautifully Cicero in his Cato, or essay on old age, illustrates this sentiment in speaking of the soul:—"Her native seat is in heaven, and it is with reluctance she is forced down from those celestial mansions into these lower regions where all is foreign and repugnant to her divine nature. But the gods, I am persuaded, have thus widely disseminated immortal spirits and clothed them with human bodies, that there might be a race of intelligent creatures, not only to have dominion over this our earth, but to contemplate the host of heaven, and imitate in their moral conduct the same beautiful order and uniformity so conspicuous in those splendid orbs;" and also wherein he quotes, on the authority of Xenophon, the dying expression of Cyrus—"For my own part," declared that great man, "I never could be persuaded that the soul could be properly said to live while it remained in this mortal body, or that it ceased to live when death dissolves the vital union." Socrates, Plato, and other great men of antiquity, whose opinions have been handed down to our day, expressed similar ideas. Thus religious convictions fill a void in the human mind, but the confidence with which they are entertained is inspired by different considerations from any which would govern the belief on other subjects, because the same evidence is unattainable; yet there must be confidence to insure enjoyment. Truth, with respect to all the affairs of life, is that at which intelligent beings aim, and where it can be attained, none but perverted understandings are satisfied with any thing short of it; and that study or science which tends to deduce the truth with respect to what most directly concerns life, property, the promotion of happiness, and the alleviation of misery, might well claim man's respect and national care. Such is the aim and tendency of statistics. Whatever tends to prolong life not only serves to extend the most precious boon, but promotes the enjoyment of its entire duration, for whatever conduces to the healtful preservation of the human system to that period when nature imperatively demands her own, contributes to relieve us from life's besetting ills and disquietudes, and renders us the more susceptible of real enjoyment and of fulfilling the design of our creation.

While statistics do not profess to heal diseases of the body or the mind, the unerring certainty with which they instruct mankind of the existence of remedial causes, leads to the investigation of their origin, and lays open defects, which, when exposed, may be said to be in a fair way to be abated or remedied. To statistics are, we indebted not only for the knowledge of such a result, but for the fact itself, remarkable as it is, that "the same number of persons which in the early part of the eighteenth century produced one hundred and six deaths in England, furnished only sixty-six deaths per annum at the commencement of the present century, and twenty years later only sixty-two," reducing the mortality two-fifths in the space of a century. Statistics develop the real condition of the human family, taking the place of vague and unwarrantable ideas whereon absurd theories were established, and which, by diverting the mind from truths, and wasting the intellectual energies of man upon illusions, threw discredit upon knowledge itself. Their object is the amelioration of man's condition by the exhibition of facts whereby the administrative powers are guided and controlled by the lights of reason, and the impulses of humanity impelled to throb in the right direction. Viewed in its application to every man personally, or to the amelioration of the condition of the human family, this science presents the strongest claims to our consideration, and on the principle involved in the admirable declaration of Cremes to Menidemus in the drama of Terence—Homo sum; humani nihil a me alienum puto—this study commends itself to the attention of every Christian and philanthropist.

We may seem to make slow progress in developing the utility of statistics as the results flow from our social or individual efforts, but we can afford to pursue the slow and laborious process of experimental investigation and mathematical deduction, and patiently witness the apparent triumph of short-lived theories, in view of the certainty and utility of our results.

Although moral statistics are of comparatively recent origin, they have accomplished more in the last half century for the alleviation of misery, the prolongation of life, and the elevation of humanity, than all other agencies combined—they are the practical workings of an elevated Christianity.

Statistics have been termed the science of social facts expressed by numerical terms; their object the knowledge of society, considered in its elements, economy, condition, and movements.

The basis of Political Economy, the contributor to Geography, and indispensable to History, they present this marked difference, that while the battles, triumphs, and conquests of the latter are its most attractive features, the former best fulfil their mission under the blessings of peace. The mere numbering of the people is an institution of great antiquity, whereof we have the first example recorded in Holy Writ, in the enumeration of the Israelites, while, through all ages, except when barbarism prevailed, history has left examples of such investigations, very different, however, in character and aims from some which distinguish the nineteenth century. Although occasional resource has been had to statistics by all civilized nations when peace permitted and prosperity seemed to recommend it, we find they have often had to struggle for an existence, and not unfrequently had their revelations altogether obliterated and their developments smothered when their truths indicated mal-administration or registered decay. Thus we learn from history that the alarming diminution of the French population, occasioned by the revocation of the edict of Nantz, was purposely concealed, and that the statistics of France instituted by Louis XIV., after the treaty of Riswick, and fostered by such men as Colbert and Vauban, were abandoned after the disasters of Hochstadt and Ramillies, only to struggle feebly into life and immediately to die, until deemed necessary by the first emperor to register with figures the greatness of his conquests and the success of his arms, to find a grave with his misfortunes, only reviving again in 1833, never, we hope, to be again obliterated.

With these preliminary remarks, I will, for a little time, endeavor to occupy your attention with some account of the statistical investigations of different peoples, with a view to apprehending the cotemporaneous progress of Statistics.

Among the Greeks, Romans, and Egyptians, the enumeration of the people was no unusual thing. By reference to Dionysius of Halicarnassus, Book I., c. 84, it will be seen that he establishes the data of the irruption of the Gauls, in which the city of Rome was taken, and from that the date of Romulus, which he says appears "particularly by the records of the censors, which were transmitted with great care, in which I find that the year before the taking of the city, there was a census of the Roman people, to which there is affixed the data, which is this—'In the consulship of Lucius Valerius Potitus and Titus Manlius Capitolinus, the 119th year after the expulsion of the king.'" The same author, in his fourth Book, ch. xv., describes the institution of a festival termed *Pagnalia,* to which the entire population contributed each a certain piece of money, differing in amount as to men, women, and children, and of which he says, "when these pieces were told by those who presided at the sacrifices, the number of people, distinguished by their age and sex, became known." Not only did Servius Tullius establish this ingenious method of at once improving the treasury and taking the census of the living population, but he instituted a plan for ascertaining the number of births and deaths, and the time at which every male person arrived at the years of manhood. In his history of Laertius, the first dictator, he refers to the revival of this, "the wisest of all the institutions established by Servius Tullius, the most popular king," in order "to register the valuation of their fortunes, adding the names of their wives, with the names and ages of their children, and their own age," Book V., c. lxxv. The last lines of this renowned writer which have reached posterity, relate to this institution, and the eightieth chapter of his eleventh book abruptly ends while describing the demands made to the senate for the re-establishment of "particularly the most necessary of all, the custom relating to the census," which had been disused "for seventeen years since the consulship of Lucius Cornelius and Quintus Fabius." Livy, in the first book of his History of Rome, bears testimony to the importance of the census as established by Tullius, which he terms "an ordinance of the most salutary consequences in our empire." This author also refers to the census combining a survey and description of all the lands and houses, and the entire revenue of the Roman people, (B. C. 440.) In his twelfth book the same author refers to the survey of the twelve colonies presented to the senate by the censors. Tacitus mentions a census in the handwriting of Augustus, which contained an exact account of his dominions. The learned French statist, Moreau de Jonnes, gives the details of the Roman population for thirty-six censuses, extending through a period of near eight hundred years, from the time of Servius Tullius to that of Vespasian, and remarks that "no people of modern Europe presents an example of arrangements so numerous as those made at Rome from the earliest times, to ascertain by periodical enumerations all the details of her population, and the changes that occurred in its condition from one year to another."

In the first book of Cæsar's Commentaries (chap. 21) we learn that when that general seized the Helvetian camp, he found therein a census by names, wherein was recorded the warriors, the old men and women, and children, which is the most ancient statistical record to which reference is made of any portion of the country now embraced within the French empire, and it stands out prominent and solitary from any thing similar with reference to that country for many ages. With the Roman census you are all familiar; its object is plainly indicated in the second chapter of Luke, and seems to have led to the fulfilment of prophecy respecting the birth-place of our Saviour. In the remarkable discourse of Xenophon upon the manner of increasing the revenues of Athens, we find everywhere inculcated the admirable maxim that the true wealth and greatness of a nation consist in the numbers of people well employed. The aim of this remarkable political essay was to demonstrate the feasibility of supporting the population of the state by the development of their native riches, rather than by the oppression of their neighbors by tribute and taxes. It seems a well-authenticated fact that the census of China was taken nearly twenty centuries before our era, if any reliance is to placed in their most venerated writings, which state that it was engraved on their public monuments in order to its preservation, and to prevent any alteration of its text.

The Arabs, distinguished as they were in all relating to figures, and who possessed a genius for calculation, were careful cultivators of statistics; and having conquered Spain in the eighth century, we learn that they very soon obtained an exact account of the country, its cities, population, and revenues.

While thus it is evident that the nations of the Old World appreciated the value and importance of statistics, it appears from undoubted authority, that on the discovery of this continent people were found who were not ignorant of their utility, but who, with all the disadvantages of an unwritten language, maintained copious statistical records.

The ancient Peruvians were found to possess accurate accounts of their numbers, wealth, and ages, even to those at the breast, and their social condition, by means of threads of different colors and strands, peculiarly knotted, "whereby they could express the greatest number at which arithmetic could arrive," the details whereof may be found in the History of Garcilasso de Vega, Book VI., c. 8. The Mexicans, according to Herrera, were found to be but little, if any thing, behind the Peruvians. Under the reign of feudalism and ecclesiastical authority many centuries elapsed during which statistical operations seem to have been entirely suspended, and all traces of their existence in Europe are lost until they were within a recent period reproduced by the more benign institutions of modern days, when we find modern civilized nations imitating the ancients, of which, perhaps, the earliest illustration is afforded by that venerable record yet extant, termed *Doomsday-book*, made in the time of William the Conqueror. This record, the work of eight consecutive years, was conducted somewhat on the plan of the ancient Roman census, except that it was made from the personal visitation of commissioners sent into every county and shire, by whom juries were summoned in each hundred, who obtained from the people, on oath, the name of every manor, of its owner, extent, and the quantity of wood, pasture, and meadow land, the number of ploughs, mill-ponds, and fisheries, its capability of improvement and prospective increase of valuation, with the slaves and live stock. These investigations, however, had for their aim the safety and good of the state, and the primary object was to ascertain, not improve the condition of the people further than to promote harmony by an equalization of their burdens in augmenting the revenues of the State.

In the nature of things such investigations, occur in whatever age they might, could not fail of ends beneficial to the people at large, or exist without being accompanied by the correction of abuses. An early illustration of this occurs in the saying with respect to Cato the Censor, repeated by Plutarch, that he was not less serviceable to the Republic of Rome, by making war against immorality, than Scipio by his victories over her enemies, and the setting up of his statue in the temple of the Goddess of Health, under which was placed an inscription, not of his triumphs in war, but one signifying that it was that of Cato the Censor, who by his good discipline and ordinances reclaimed the Roman commonwealth, when it was declining and falling precipitately into vice.

The scope of national statistical investigations has been gradually enlarging in some countries, while in others, Spain for example, no advance has been made beyond irregularly and imperfectly numbering the inhabitants; but it is only of recent date that these inquiries have been systematically prosecuted with the benevolent view of improving the general condition of society, or the amelioration of the misfortunes of those who are morally or physically unable to assist themselves, or of making known their wants.

It is only now, and that to a somewhat limited extent, that the mission of statistics develops itself in carrying out the example of our Saviour while on earth, and by means of those agencies within man's control, conferring the greatest boons upon suffering humanity, which He wrought by miracle. Doubtless the object of the first statistics was to ascertain the amount of destructive or repelling power, after which followed a more general enumeration of the people, succeeded in the progress of discovery by geographical and commercial statistics, followed by those of mining, agriculture, and manufactures; while now its errand embraces the march of the mental faculties and progress of knowledge, combining all the elements of human welfare, charity, benevolence, and religion. The blind are taught to read; the deaf and dumb to express themselves, while sound mental faculties are restored to the insane; a thousand rills of mercy are put in motion, and the heart of man softened by the misery which is brought to light, appealing irresistibly to his humanity, benevolence is forced to flow in rivers of beneficence, whereby man is made happier and better. Of this your own knowledge will suggest innumerable illustrations.

While it is my particular object to address you mainly with reference to national efforts evincing an interest in statistical investigations, with some history of their operation, I cannot entirely pass over some of the individuals who at different times have manifested a high order of talents, and exhibited great perseverance and energy in their efforts to ascertain and make known the social condition of the world; for to them is due not only much of the value of history, but the success of the science itself. Were we to attempt even the mention of all who, by their writings and opinions, have declared their consciousness of the importance of such investigations, we would be compelled to omit every thing else, as such a list would comprise a long array of the greatest names which adorn history. When we read their works our sympathies are moved by the appreciation they entertained of the surrounding darkness, and the difficulties they realized of presenting any reliable basis whereon to construct their deductions, while we cannot but admire the fortitude and genius displayed in their investigations as well as in the results of their calculations. Omitting further allusion to the men of antiquity, since the existence of whom all traces of statistical studies were lost for centuries, as illustrative of the origin of these pursuits in more modern days, and the labor attendant upon these studies in the absence of such reliable data as we now possess, we may refer to one of the numerous works of Sir William Petty, a member of the Royal Society, and founder of the great house of Lansdowne, certainly one of the earliest statistical writers in our language, dated 1686, whereby will be perceived the equally singular data from which he calculated the population of London in his day—thence the population of England, and the estimated population of both in 1840, by which he determined that of the city at about four times what it has attained, and that of the country at only one-fourth its real numbers. (Sir Wm. Petty's *Essays concerning the multiplication of mankind.* London, 1686, pp. 20, 21.) By the examination of such works, we perceive the embarrassment which surrounded the statist of the seventeenth century, and the difficulties at that time experienced by a man of rare genius in determining proximately the first element of political economy, the number of the people, and how, from a few deaths, he arrives at the number of the living, establishes the population of London, whence he deduces that of the kingdom.

In the absence of better data his method was the most rational, and his view of the relative rates of city and country mortality, I imagine, would not be very far from true, if applied at the present day to this city. Adam Smith, in his Wealth of Nations, makes frequent mention of the labors of this ingenious political economist, who could educe light from darkness, and to him attributes the first reliable estimate of the Irish population which formed the basis of subsequent estimates for near two centuries, for it was not until 1821 that a reliable census of Ireland was made by order of the government; and that of England itself goes back no further than the present century, and I regret to say that in the variety of its details it has experienced but little improvement, though admitted now to be taken with accuracy.

Perhaps the greatest modern statistical work resulting from individual effort, and combining geographical and statistical knowledge in the most pleasing combination, is the statistical account of Scotland by Sir John Sinclair—a work embracing contributions from nine hundred pens, and filling twenty volumes. This distinguished statist manifested a correct view of the importance of the science he loved so well, no less by his great and imperishable work than the declaration that it was the most important of all sciences, and, in preference to any other, ought to be held in reverence. "No other," he affirms, "can furnish to any mind capable of receiving useful information, so much real entertainment, none can yield such important hints for the improvement of agriculture, for the extension of commercial history, for regulating the conduct of individuals, or promote the general happiness of the species"—a tribute to statistical knowledge, in comprehensiveness and beauty, not a whit behind the oft-quoted tribute to general science by Cicero, in his oration *pro Archia poeta.* Parish registers existed in some of the continental cities, before they were used in England, where, only within little more than a century, have the ages and causes of death been recorded. The first "Bills of Mortality" in England originated at the close of the sixteenth century, to quiet the consternation produced by the frightful ravages of the plague, "which were found so useful as to be continued from the 29th December, 1603, to the present time," while the first life-tables in England were the work of the illustrious Halley.

The labors of the early statistical writers of England, such as Graunt, Petty, Halley, Simpson, and others, throw much light on the subject of population and statistical knowledge in infancy, while those of Price, Smith, Porter, Redgrave, Fonblanque, Neison, Farr, Fletcher, M'Cullough, Brown, and others, have done all for statistical progress, in more modern days, which the human intellect could accomplish in the absence of those national efforts which at the present day characterize the census of the United States, France, and Belgium. They have effected more, with all their disadvantages, than we, who in some respects have better data. Their system of registration, and the facility of insight into their criminal statistics, have afforded many advantages, of which they have made judicious use. The British Government contributes all it feels the power to exercise, for the development of their social condition, while admitting its impotence to concede more. A different government financial policy, which time must inaugurate, will alone effect a change in the reserve of their people into a state of communicative confidence such as we here experience, till when, we may look in vain for much additional light on the social condition of the inhabitants of Great Britain.

In France, of late years, much has been done for statistical knowledge, but considering how the sciences generally have been cultivated and encouraged by that people, and in view of the fact that Louis XIV. made strenuous efforts for the permanent establishment of statistics in connection with the administration of government, it seems remarkable that so little should have been accomplished until within a very recent period. In 1784, Necker figured the population of France, which he calculated upon the estimated number of births, as Petty had that of England upon the deaths—and no less than three fruitless efforts were made to ascertain the productive capacities of the kingdom. At one time it was attempted to deduce the quantity of agricultural products from the number of ploughs, as was done by Lavoisier. Then, by assuming that the result in 30,000 communes might be predicated upon the facts ascertained respecting a few, as was attempted by Chaptal, or by ascertaining the products of a square league and therefrom deducing the agriculture of the kingdom, as was done by Vauban. In that government statistics have been doomed to trying vicissitudes, arising from the difficulty of their execution, but perhaps more on account of the unpalatable revelations which, in deference to its mission and to truth, it was compelled to make concerning the state of the kingdom. In fact, the fear of popular opposition frequently induced the agents of the French government to resort to artificial calculations to obtain figures which an actual enumeration alone could ensure. In 1822, it was provided by a royal decree, that a general census should be taken every five years; nevertheless, we learn that in place of the actual enumeration, which should have been made in 1827, the number of inhabitants was declared by simply adding to the population of 1822 the excess of births over deaths for the intermediate time, and the result was determined by royal ordinance to be authentic.

French writers claim that a census of France was made in the sixteenth century, under Charles IX., of which they profess to give the results (although no traces of the original are to be found) at 20,000,000 of population, whereas, by that published in 1720, their numbers only amounted to 19,500,000. It is not impossible that both statements are near the truth, in which case we learn the depopulating effects of war and oppressive edicts. Among the distinguished men of the past century, who adorned statistics by their writings, may be mentioned M. Buffon the celebrated naturalist, and Pascal the philosopher and Christian—both ornaments to science and to France.

In 1833, statistics in France were established, we hope,

on a firm foundation, since which they have received the fostering care of government and the immediate attention of some of the first scientific men of the empire, to whose discretion their execution has been entrusted—and for the last fifteen years volume after volume has appeared, manifesting, on the part of that government, a determination to take the lead in every branch of statistical investigation calculated to illustrate the moral, social, and material progress of their people.

To the labors of the polite and indefatigable Moreau de Jonnes, long the chief of the French bureau of statistics, have I been indebted for many of the statistical facts presented in this paper, in connection with the mention of whom I may state that he expressed to me personally, a few years since, his high appreciation of our system of enumeration, declaring at the same time that our schedules for obtaining the primary facts are the most perfect and comprehensive ever used.

In Russia, that wise monarch, Peter the Great, established, in 1722, the registration of births, marriages, and deaths, and in the year succeeding, caused the first enumeration of all the people to be made, prescribing its renewal every twenty years; measures which have since been conducted with regularity and care, so that the Russian government possess what no other government can boast, the statistics of the movement of her population for a period embracing considerably more than an entire century. The long period through which the Russian government has ever successfully pursued this great work, is a triumphant vindication of its utility, and a sufficient answer to the allegations continually made by uninformed persons, that but little or no reliance can be placed on the revelations of a national census; and when we look upon the 60,000,000 of people to be enumerated, and the immense extent and character of the country to be included, we may well admire the enterprise, perseverance, and capacity of those intrusted with so responsible an undertaking.

Sweden, too, for more than a hundred years, has prosecuted statistical investigations, and on recommendations originating with the Academy of Sciences at Stockholm, which counted the celebrated Linnæus among its members, established a system of enumerations and registrations which have been conducted with care, minuteness, and perfection worthy their academic origin. To Sweden is the world indebted for the invention of mortality-tables and the illustration of their uses.

In Prussia, statistical science which owes its establishment to the penetration of Frederick the Great, has long received the fostering care of government, but it was not until 1806, under the reign of Frederick William III., that a bureau of statistics was established at Berlin. After near half a century of independent, successful prosecution of statistical investigations, including a triennial census enumerating the population, domestic animals, mines, and manufacturing establishments subject to taxation, and the registration of births, marriages, and deaths, this government formed a union of the thirty-nine principal German States into a single commercial association termed the Zollverein, whereof there are conducted special and detailed statistical labors, which, until his death, were under the direction of Dieterici. In referring to this union, Moreau de Jonnes remarks, that "this is the first time that, in Europe, different countries leagued themselves under the banner of science, to produce a work on social economy, expressed by numerical terms, and in which each participated with a fraternal spirit."

For about thirty years the statistics of Belgium have been prosecuted, irregularly enumerating the inhabitants, to the year 1846, when, as an independent State, their first census was taken.

In the report of this work, drawn up by Quetelet, and presented by the Minister of the Interior to the King, we perceive the exhibition of wonderful hesitation and care in the commencement and execution of this their first effort on an extended scale, to obtain the general statistics of the kingdom, a duty which was intrusted to a Board of fifteen persons, embracing some of the most distinguished savans of Belgium, including Quetelet, Detournay, Ductipaux, General Trumper, Visscers, and Heuschling.

The census of Belgium, to which I allude, embraces at once population, agriculture, and industry, with many details of great interest, while those previously taken extended to population alone. The formation of agricultural and manufacturing statistics presents Belgium in a new aspect. For the first time the situation of the productive wealth of the country, in its totality and details, is authenticated; it is also the first time that statistical science has there been called on to proceed to so extended and complicated an undertaking, which it has successfully accomplished. It would appear that the statistical Board, before venturing upon their great enterprise of the national census upon an enlarged system, made an experiment as to the population of the city of Brussels. The history of this work is narrated by Baron Quetelet with great clearness. In fact, it was my intention to have read a portion of this complete, yet simple, history of the experience of the agents of a cautious, prudent, and enlightened government in one of its measures of the greatest public utility, which I thought would not prove void of interest to an association like this, as it inaugurated a new era in the history of their statistical progress, and in the pursuit of which we may discover, as its results compared with our own, the influence of different institutions upon people of very kindred pursuits, of similar mental progress, but geographically wide apart—but time will not permit. The prosecution of such a work, only second in detail to that of our own country, illustrates an advance in Europe which the world might not have witnessed for a century, but for the success which has so long attended the prosecution of our own. The Report is occupied in detailing the groundwork and results, whereby it forcibly illustrates the value of their labors in the correction which they effect of many false or erroneous ideas previously entertained, on many questions relating to the condition of the kingdom.

It is about a century since, that the Emperor of Austria, Francis I., decreed the census of his kingdom, but it is only within the present century that regular statistical operations are conducted to the extent of numbering the people. Among modern statistical writers on the continent, in addition to those already mentioned, appear the names of D'Alembert, Euler, Lagrange, Laplace, and Condorcet, "all throwing a charm over a science with which their names have been associated."

Lastly, in noticing the cotemporaneous progress of statistics, we come to consider what has been done in our own land. A French statistical writer, in speaking of our country, utters the following compliment: "The United States," he remarks, "present in their history a phenomenon which has no other example. It is that of a people which begins the statistics of its country on the day on which it lays the foundation of its social condition, and which regulates, in the same act, the enumeration of its fellow-citizens, their civil and political rights, and the future destinies of the country." In another place, referring to the penalties imposed for a refusal to answer the interrogatories of the Marshal in his lawful duty, he says, "Statistics were treated seriously eighty years ago by a people that, however jealous it is of its liberty, does not hesitate to punish, as a culpable infraction, what is elsewhere regarded as an action of no consequence, or treated with futile opposition. In the United States," he continues, "it is a civil duty, the importance of which seemed so great to the convention over which Washington presided, and which numbered among its members a Madison, a Livingston, and a Franklin, that it imposed fines upon a citizen or a magistrate who disregarded it."

I could present no better introduction to a short historical account of our national censuses, the first of which was taken in 1790, in accordance with the second section of the first article of our Constitution, which requires an enumeration of the inhabitants within each subsequent decade, in pursuance of which, we have made seven enumerations of our people, and have commenced preparatory measures for taking the eighth. The provision of the constitution requires nothing more than a mere numbering of the inhabitants, made necessary for an equal representation in Congress, and the apportionment of taxes (should it become necessary for the support of the government) among the several States in proportion to the population of each.

Dr. Davenant, a powerful writer of the seventeenth century, commences his volume of discourses on the public revenues and trade, printed in 1698, with the declaration, that "He who advances a new matter, is bound to show the foundation he builds upon, whereby the public may better judge whether he be right or wrong in his superstructure," upon which principle I will, without troubling you with an account in detail of the plans adopted for the several intermediate enumerations which varied with each census, endeavor to present a short account of the method adopted in taking the seventh, which is rendered more important from the fact that, in accepting the plan for this work, Congress, by statute, made it the basis for subsequent enumerations, unless a law changing its nature should be passed previous to the first day of January of any year wherein the enumeration under the Constitution was to be made; a contingency not at all likely to occur so as to prevent the taking of two national censuses on the same plan, if we may judge from the weak support accorded in the last Congress to propositions introduced for the purpose of effecting some modification of the law, and the short period which now intervenes previous to that indicated in the statute when the time for amendment will have passed. The wisdom of this provision will readily be conceded by all who have observed the uncertainty attending legislation with reference to our national census, especially the doubt which seemed to attend for a long time the enactment of the necessary law for prosecuting the seventh enumeration, which was passed on the last day of the session. It was the danger which ever seemed to exist when this subject was agitated in Congress, of a failure to make provision for its timely execution, that induced Mr. Vinton, of Ohio, to insert the amendment, which being adopted, forever precludes the failure of the decennial census, unless the law should be repealed. To the sagacity of the same gentleman is the country indebted for another very important and tranquilizing amendment to the bill, namely, that of establishing the ratio of Representation in the House of Representatives, upon a just principle, in advance of a knowledge of any local effect it might have, and thus taking out of that body one of the most dangerous and exciting topics of legislation. By this latter amendment the whole number of representatives is fixed at two hundred and thirty-three, who are apportioned among the States respectively by the Secretary of the Interior, by dividing the number of the free population of the States, to which, in slave-holding States, three-fifths of the slaves is added, by the number two hundred and thirty-three, and the product of such division (rejecting all fractions of a unit) shall be the ratio of representation of the several States; but as the number and amount of the fractions among so many dividends would, of course, in the aggregate, be sufficient to reduce the number of Representatives below the number specified, it was provided that the whole number should be supplied by assigning to so many States having the largest fractions, an additional member each for its fraction, until the number of two hundred and thirty-three members should be assigned to the several States. It is also provided that new States being admitted subsequent to any one of the decennial enumerations, shall have representatives on the same basis, while it is at the same time provided, that such excess in the number of members of the House of Representatives shall only continue until the next apportionment of Representatives under the next succeeding census. The number of Representatives being fixed by law at two hundred and thirty-three, it may be asked how it happened that, under the last apportionment, the number of Representatives was two hundred and thirty-four? to which I reply, that a portion of the returns of California having been destroyed by fire at San Francisco, enough only reaching the Department to entitle that State to one member, while it was evident she possessed population enough for two, and the Representatives from South Carolina, contending against being deprived of the claim she rightfully possessed for the eleventh member admitted by virtue of the fractions ; the matter was compromised by a law of July 30, 1852, whereby to California was conceded two members, (with which number she had been admitted into the Union,) and the number of Representatives was at the same time extended to two hundred and thirty-four, until a new apportionment should be made under the eighth census ; and in the same law provision was made against future like contingencies. The practical execution of the apportionment was as follows: The total free population of the States amounting to 19,846,710, and the slave population to 3,200,380, whereof three-fifths being added to the former made a representative population of 21,766,938; this being divided by 233 the product became 93,420, and the representative population of each State being divided by the latter sum, the quotient expressed the number of Representatives. This process having been completed, whereby two hundred and twenty-two members were apportioned ; in order to make the complement eleven members were to be added, one to each of the eleven States whose population presented the largest fractions. Thus was settled, at least for a time, what has been viewed as one of the most dangerous and exciting topics connected with our national Administration, to the disappointment, I am inclined to think, of some who were not without the hope of Congress being unable always to meet this constitutional requirement, whereby confusion and disastrous consequences might have followed. Thanks to a superintending Providence, and for our consolation, there seem ever to be found in all emergencies, men whose calmness, judgment, and discretion, may be relied on to ward off impending evils and rescue our ark from danger.

In the preliminary arrangements for the seventh census, unusual provision was made for securing the best plan upon which to conduct its operations; and although some of the details adopted by the Census Board and presented to Congress, met, as might have been expected, strenuous opposition from a few members in both houses and of some persons without, the hostility was not as great as was anticipated, and the recommendations of the Board were adopted by a large majority with no other than slight verbal alterations. As "prosperity only crowns the efforts of industry when its operations involve a judicious division of labor," so the successful investigations of statistical science depend not only upon specializing each branch of its inquiry, but, following up every ramification to its elementary beginnings, from which follows the impossibility of compiling an accurate or reliable statistical account of any people without first obtaining the history and condition of families and individuals, a fact which although lost sight of in practice, through intermediate centuries, was as well understood when the children of Israel were numbered in the wilderness of Sinai, and the Roman people responded to the questions of the censors, as now. This consideration, long lost sight of, though supported by such historic examples, induced the Census Board to prepare a series of tables of great simplicity and scope, the records whereof embody, for the time, the history of our entire population, comprising one great family record which will be appealed to for generations to come, as they now are, for details which can be found nowhere else.

The first table relates exclusively to the free population, and was prepared with a view to obtaining, in the simplest form, the greatest number of facts essential to an understanding of the numbers and condition of the American people in all their relations, together with the number of habitations. By this schedule we have the number of families—the age, sex, and color, distinguishing between the white, black, and mixed, in order, if possible, to obtain data whereby we might form some judgment respecting the admixture of races upon their physical condition; the nativity of each inhabitant is returned, with the number of those married within the year; those attending school, with the adults unable to read and write, the deaf and dumb, blind, insane, and idiotic, paupers, criminals, and convicts—and all the facts legitimately applying to the history of each and every individual are given together, so as to present a most interesting combination of facts, which exhibit the relation of age, knowledge, crime, physical or mental infirmity to all of every sex, color, condition, and origin, which will be taken with more and greater accuracy, as each effort shall furnish the means to remedy the causes which have led to imperfect returns. Many of these details are rendered the more necessary in our country because of the want of State laws for the registration of deaths, births, and marriages; and although the advantages arising from the decennial enumeration of these facts may seem comparatively limited, they furnish in the absence of data more general in their application, a great deal from which we may be enabled to form opinions, and institute comparisons; eventually, perhaps, from their decided utility, inducing the several States to enact and enforce laws for the registration of the three great epochs in man's history, of the importance of which several have already manifested their appreciation.

The second schedule gives the numbers and ages of all the slaves with their sex and color, together with the number who have escaped from bondage—the number manumitted, with the deaf and dumb, the blind, insane, and idiotic. It was not deemed necessary to incumber this schedule with many of the features of the first, because the profession, occupation, or trade of a slave, was a matter optional with his master, and varied with circumstances ; he could not own real estate ; his birthplace could not generally be recorded with certainty ; their marriages were a matter of doubt at the best; Congress, doubtless, acted wisely in declining to institute investigations, which would be looked upon as a stretch of authority, lead to no practical good, make the census unpopular, and defeat the obtainment of many of the interesting and important details which we are now able to acquire.

The third schedule relates to the statistics of mortality, by which it was endeavored to obtain for the first time a reliable account of the number of persons who died within the year previous to the first day of June, the day with respect to which the enumeration was made. The facts ascertained, are the name, age, sex, color, civil condition, marital relation, nativity, occupation, disease, and duration of illness, of all persons dying within the previous twelve months, in each State and Territory. In determining to procure these statistics, much opposition was experienced in and out of Congress, from persons who pretended to foretell that such information could not be obtained with any approach to accuracy by means of the census.

These predictions were not verified by the results. For the country in general, they are believed to be quite accurate, for a first effort, while for cities, where great fluctuation of population occurs annually, and where an important event may occur in a family and sooner be lost sight of in the excitement of places where a dense population exists than in the quiet of country and village life, the returns are admitted to be often materially defective ; but fortunately, as a general rule, where the results we aim at are most difficult of acquirement by means of the census, there exist other means of obtaining them, of which we fail not to avail ourselves. Admitting the returns to have been imperfect, they are not without great and important uses, for it does not follow that imperfections destroy the value of statistics, unless they are the result of design or are of a nature to present exaggerated results and lead to false deductions. In a letter of Quetelet to the Grand Duke of Saxe-Coburg and Gotha, in answer to the question, "Can any advantage be derived from incomplete statistical documents?" that distinguished philosopher remarks as follows: "The number of the population is, without doubt, the most important statistical element. However, I do not think there exists a single country in the world in which this element is well known, and in this I speak not of mathematical precision. In this state of things, should we reject all the results into which the number of population enters although evidently faulty? I think not, especially when we would consider relative quantities rather than absolute." No opinion of mine could add weight to such authority—hence, from these numbers we arrive at the proportion of diseases of a particular character and diverse natures which occur in various situations, and obtain a combination of facts illustrating the influence of climate upon all manner and descriptions of people who live and die within our borders; and continuously pursuing and improving the mode of execution, it will not be long before their effect will be better understood and studied by the emigrant, in determining his future home, and the citizen in choosing his place of residence. Exhibiting the effect of occupation and trade, they will not be without their influence in deciding the choice of a profession, while dwellers in cities and large towns will learn the necessity of better sanitary regulations and the advantage of a purer atmosphere at home, or the necessity of a country residence for their own benefit and the security of their offspring.

Were it part of my object, this evening, to occupy your attention with an account in detail, of the ascertained results of statistical investigations, and the benefits which are known to have followed, rather than their history, it would not be difficult to make manifest their value as illustrated by developments in Europe and America, declarative of the extraordinary fatal contingencies associated with life in densely populated and illy ventilated towns, and the average prolongation of days promoted by remedial appliances for improving their sanitary condition. It would appear from well-established facts, that every ray of sunlight, and each additional breeze admitted from a pure and healthy atmosphere, mingled with electric rapidity with the blood of all the people, and with healing on their wings brought life, animation, and courage. Fortunately, perhaps, for our sympathies, these causes which are preventable, and effects which are hidden because slow, wide-spread, and gradual, are not apparent at any one period, or disclosed in any one portion of the year, unless perchance developed by pestilence. Were the preventable deaths which annually occur in this city of New York, to take place in any one week, they would throw a pall over every street and lane, and strike such a terror into the minds of the people as to paralyze industry and stagnate trade. Now, apart from the consideration involved in the great amount of preventable *sickness*, (not unto death,) whereby such a vast amount of labor and earnings are lost and misery inflicted, were the amount of preventable annual mortality for two consecutive years, to be concentrated into the same month for each year, you would feel no hesitation in proceeding at once, airy and well ventilated as your city may seem, to demolish squares of buildings to give place to public grounds, while beautiful parks would appear as if by magic, throughout the length and breadth of this great metropolis; and while you would effect this with a hearty good will, regardless of the cost, it would be found a wise and economical expenditure of treasure. The fall of population, like the dropping of the leaf of the evergreen, is not realized because of so silent, slow, and gradual occurrence, and we look upon what is left and what comes on to take the place of the departed, quite unthoughtful of the silent, deadly, and inevitable influences at work. To these causes and effects, statistics open the eye of reason, alas! too often impotent to provide a remedy, or indifferent to its exercise. Our first efforts to obtain the statistics of mortality have not been without their fruits even at a very distant point from where they were collected together, and that even long before their publication by government. In this I refer to the revelations made by Dr. E. H. Barton of New Orleans, the author of two valuable contributions to medical and statistical science—one a report of the preventable diseases of Louisiana, Mississippi, Arkansas, and Texas, accompanied by maps, and the other a large volume on cholera. The first work forcibly illustrates the value of our mortuary statistics, and the preparation of such a volume, by Dr. Barton, is refutation sufficient of any charge of the uselessness or unreliability of these statistics. In pursuance of a wish expressed by that estimable man, he was furnished, in 1851, with such developments respecting the mortuary returns of these States as the revelations of the census admitted, and I received from him in April, 1852, a letter communicating an account of the uses to which he had applied them, which I will read—before doing which it may not be out of place to state that his opinion of the value of these statistics, as we ascertained them, was fully endorsed by some of the most eminent medical men of the country:

NEW ORLEANS, *April* 27, 1852.

MY DEAR SIR: I have just finished the task, the materials for which were so kindly furnished me by you, for which be pleased again to accept my sincere acknowledgments.

I am of opinion the great work you are now laboring to accomplish is second to none in importance, undertaken by our, or indeed any other country, and will be more appreciated just as it is understood.

Out of the medical returns, I have constructed a Sanitary Map for each of the States of Louisiana, Mississippi, Arkansas, and Texas, to accompany and illustrate my report to the American Medical Association, meeting in Richmond, the first Monday in May. I beg you to accept a copy of each of these maps, in consideration of the favor you have done me. The only explanation required, probably is, that as it was impossible to express on one series of maps, the diversified sanitary condition arising from the greater or less prevalence of every form of morbid action, I have selected that class which embraces the mass of those maladies which constitute the distinction between healthy and sickly communities, so as to compare one country, State, or county with another, and which are more or less under the control of police regulations and sanitary laws; the Zymotic, (including fevers, cholera, intestinal diseases, &c.)

I have arranged them, as per marginal explanation, according to the number per thousand of the population which have suffered from them.

You see Louisiana has there a distinguished showing; this was owing to the great prevalence of cholera, (25 per cent. of the entire mortality having been caused by that disease.) You see how it followed the water courses, and the great routes of human intercourse, and how free the seacoast and pine woods were from it.

I wish the whole country was thus delineated; your labors would be then appreciated, and we should be enabled the more easily to compare by the next census—showing, with this, how our countrymen may have been benefited by the lesson; for, as I said above, these diseases arise in the main from *preventable conditions*, such as an advanced civilization should correct ; and, to use the language of Dr. Rush, such as should be enforced by a *penalty!*

I have added some tables comparing the rural districts of my four States with the rural districts of Cuba, and the cities of Havana, Mexico, &c., from details I collected while travelling in those countries when in the army. They are curious, and highly in our favor.

I have also added all the meteorological details I could procure of the States assigned to me, and particularly of my State, and have added the *average solar radiation* for each month in the year. I remain, with high regard,

Very respectfully, your obedient servant,

To J. C. G. K. E. H. BARTON.

I received from this useful and benevolent man, in the month of September last, a communication wherein he expresses his approbation of the recommendation of the Secretary of the Interior in his last Annual Report, that the eighth census shall be taken in conformity with the law prescribing the plan of the seventh and subsequent enumerations, and dwelling at some length on the wisdom and propriety of the Secretary's reasons for desiring to conduct the coming enumeration upon the manner of the previous one ; and expressing in strong and confiding terms the pleasure with which he anticipates the mortuary returns for the materials they will furnish, whereby he can prosecute his favorite investigations, and by throwing additional light upon the causes of mortality, promote that of humanity, little dreaming, alas! that before the letter reached me he would be inscribed among the five hundred thousand, whose names will form the melancholy record of the dead comprising the mortuary catalogue of the eighth census.

The fourth schedule relates to agriculture, giving the number of farmers and planters, the quantity of improved and unimproved land, the value of farming implements and machinery, the number and description of live stock, the value of animals slaughtered ; and enumerating thirty varieties of agricultural productions which enter into consumption and contribute to the commerce and manufactures of the world. "Were we to attempt in this country," said Joseph Hume on one occasion, as he was attentively scrutinizing some of the returns in the form they came from the hands of the marshal, when he came to inspect the agricultural schedule, "were we to attempt in this country, officially to obtain what your people have so cheerfully and universally communicated in accordance with a law of Congress, it would effect a revolution in less than six weeks." This distinguished man attributed the difference to the general diffusion of intelligence among our inhabitants, the sympathy and reciprocal obligations and confidence existing between the representatives and their constituents, and the different motives which naturally would be attributed to the two governments respectively, by the people. In connection with this incident, I will mention what I look upon as a remarkable and significant fact developed in the taking of the seventh census, wherein every man was approached with a pretty formidable array of questions, which is, that in our entire country, three persons only were found to prove contumacious in refusing to answer, and that only one persisted to the extent of requiring a notification from the District Attorney of the United States, when his scruples vanished and he communicated the information demanded. I look upon this fact as important in every aspect, whether viewed as illustrative of the spirit of obedience in the people, or as indicative of the general truth of the census; for if the inhabitants generally respond to the plain questions proposed —and we have no reason to believe they responded falsely— there can be no ground for affirming that the census is not essentially reliable, if any census ever was, or any ever will be so, for there can be no more certain way of ascertaining the truth with respect to numbers than by combining the elementary details as they are given in by individuals, and in all such returns, the errors incident to human infirmity will balance each other. This truth is susceptible of proof, even in cases of conjecture where there are no data but what may seem imperfect. A judicious observer, while he might not be able to state the precise age of any gentleman here present, and therefore be without any ascertained fact upon which to predicate his conclusion, will nevertheless be able to fix with proximate accuracy, the aggregate age of this audience; and so you may take a target of the most irregular form, wherein at some point to you unknown, twenty shots had been fired, and by a little calculation determine the precise spot aimed at. How much more therefore are we justified in determining the truth of statistics based upon actual returns.

The fifth schedule relates to the manufactures of the country, is arranged on a plan of great simplicity, and constructed on entirely different principles from any ever used previously for a like purpose—consisting of fourteen columns or headings, applicable to the details of every branch of productive industry, with the number of male and female operatives, their wages, and the quantity, kind, and value of raw materials, value of fuel, the motive power, and the quantity, kind, and value of each article produced. The figures with respect to each establishment are recorded separately, so that, as is the case with the other schedules, the marshals are not required to add any two numbers together; their whole duty in this entire work being to ascertain the primitive numbers with accuracy and care, leaving every calculation to be made in the office at Washington.

The results of this portion of the census were omitted in the earlier publications of the seventh census, and were only communicated to Congress within the present year in accordance with an appropriation of the last Congress, and made public since the adjournment. As no copies of this work were printed for distribution, it may not be out of place to say something with regard to it.

Although it pleased Congress in ordering the publication of this work, to leave out much with respect to details, which, in my opinion, should have been included, nevertheless, we establish by what has appeared, some of the most important facts with respect to a people who have attained to numbers, wealth, power, and influence, with a rapidity unparalleled, at a moment when many nations of the old world appear to be rising from a lethargy wherein they seemed to rest until we should attain to the dignity of rivals, and when a spirit of emulation, such as never before prevailed, exists among all countries to render themselves independent each of the other. We exhibit the fact of a nation, the birth of which those living can remember, whereof the mechanical industry of less than one-twentieth of her people, by doubling the value of their raw materials, produced manufactured articles to the value of more than one thousand millions of dollars, giving support to one-fifth of the entire population, and promoting the commerce of the world and the manufactures of other nations by the raw materials we send abroad. It is further made to appear in what mechanical interests the capital of our people is invested, and to what industrial pursuits their energies are directed, teaching the value of those resources which enlarge our own immunity, as well as those that augment the dependence of others upon us.

In a few instances, in the details of this work, it will appear that the value of raw materials exceeds the product and the amount paid for labor. In such it must be concluded that when the census was taken, these manufactories had not been in operation a sufficient time to consume the raw materials, or that their operations were for a time suspended on account of temperature, the failure of motive power, as with machinery impelled by water, or the transient character of their objects, such as fisheries, &c. This difficulty has arisen in part by returning the monthly cost of labor, which, in the absence of specific information, was multiplied by twelve to arrive at the annual expenditure. These are exceptions which in nowise tend to impair the truth of the returns at large, or in any way misrepresent the capital and productions, the value of labor during the time of operation, or the worth of the product, and cannot in the least affect the general credibility of the work; and I have deemed them worthy of mention, lest persons unfamiliar with such contingencies might be led to form conclusions unwarranted by the facts. They are exceptions which we hope to avoid hereafter, although when understood aright, they serve rather to confirm than weaken the general reliability of the tables. In the compilation of the abstract of manufactures, I did not deem it proper to enter into any deduction of their consequences, or to record any expression of opinion as to their tendency. While there may be those to whom this reserve may appear to diminish the interest and value of the work, it was in my opinion the better course; and in view of the fact that the mere expression of immutable truths often provokes the hostility of those to whom they are unpalatable, and the difficulty of engaging in interpretation without appearing to some to have made indiscreet commentaries, or imparted a false coloring to facts, I was induced to act strictly within the spirit of the act of Congress, and, without the incorporation of any notes, "record the numbers distinct from any deduction or inference respecting their teachings, to stand in their simple integrity enduring and independent exponents of the truth." With respect to the aggregate amount and value of the several manufactures, it should be borne in mind that no interests were included, whereof the productions did not amount to the value of $500 per annum, so that our aggregates may safely be determined to be within, rather than beyond the truth by several millions of dollars. It was necessary to establish some limit, and a proper one, it is believed, was observed. The proportion of manufactures of a less sum annually than that stated, varies in different sections of the country, and cannot with our present information be satisfactorily estimated; comparatively small in cities and densely populated regions of the country, the proportion would be found not at all inconsiderable in villages and country places where living is cheap. It must not be concluded however, that the results of labor devoted to these minor interests is entirely unrepresented, as they often form some of the elements of the larger operations included in the digest.

Our sixth and last schedule relates to ten different subjects, being constructed on the same general principle with the others so as to embrace all the facts ascertainable respecting the value of property, amount of taxes, state of education, literature, religion, pauperism, crime, and wages.

While prepared to vindicate the general correctness of the seventh census, to assume that errors may not exist in details involving interests of such magnitude and variety, would be more than any one familiar with the difficulties attendant upon such investigations would venture, while a critical examination will inspire confidence in the general accuracy of the work, and prove that any serious imperfections which may be charged are more often imaginary than real, and more frequently alleged by persons of limited means of information than by statists of judgment and experience. Some claim for indulgence at least, on the part of those intrusted with the execution of the work, might, with reason, be urged on the ground of but eight days intervening between the passage of this law to take the census, and the time of its commencement, in which the instructions to the marshals were to be matured and the entire machinery of a great and important work put in motion on a new and untried principle; and with respect to the statistics of manufactures, the further fact that a period of about five years intervened from the time the superintendent ceased to be connected with the suspended work to that when he was honored with the appointment to complete it. I have thus given, in too many words, but in as few as my poor abilities would admit, an imperfect account of the schedules used in taking the seventh census, from which you will be enabled to form some idea of the variety and combination of facts they are adapted to illustrate.

As the numerous letters daily received at the census office asking information with respect to the manner of conducting the *operations of the census*, make it apparent that the details are not generally understood, I have thought a moment might be advantageously appropriated to this subject.

The General Government has in each State and Territory,

one or more judicial districts, with each of which is connected a marshal, who acts as the high sheriff in the District Court of the United States. These marshals are required by law to subdivide their districts, and for each subdivision to appoint an assistant—taking care not to include a greater population (by estimate) than 20,000 in any one subdivision.

The assistants having qualified, by oath, for the proper performance of their duties, are furnished through the marshals with blanks and instructions. In the prosecution of their work they are required to visit every house, manufactory and workshop, and when they have completed their district, are required to make two copies of their work. The original returns are filed with the clerk of the court of each county, and the copies are forwarded to the marshal, who transmits one copy to the secretary of the State for his district, and the other to the census office in Washington.

The compensation to the marshal is in proportion to the population enumerated by his assistants; should that exceed one million, he is paid one dollar for each thousand persons enumerated; should the population returned by his assistants be less than one million, he receives the sum of one dollar and twenty-five cents for each 1,000 persons returned;—a system of compensation sufficiently moderate, but which may admit of the payment of a greater amount for a lesser service, as in the case of a marshal whose returns include 950,000 at $1 25 per thousand, receives more than he whose returns do not much exceed a million, an inequality not unusual in rating fees for mileage and other services. Should the population of a district be less than 400,000, the marshal is allowed additional compensation for clerk hire.

The assistants who perform the work of enumeration are paid on a different principle, combining in a novel manner compensation for labor and travel, one which was found to operate very fairly and satisfactorily to the employees and Government. Their allowance is two cents for each person enumerated; for each farm, ten cents; for each establishment of productive industry, fifteen cents; for social statistics, two per centum on the amount allowed for enumerating the population, and two cents for each mortality return, with ten cents per mile for travelling expenses, to be ascertained by multiplying the square root of the number of dwelling houses in his district by the square root of the number of square miles in his division; the product whereof is to be deemed the number of miles travelled, and eight cents per page for the two copies.

The marshals and assistants in California, Oregon, Utah, and New Mexico, under the operation of an amendment to the law, received compensation at the discretion of the Secretary of the Interior, which was determined by the addition of one hundred per cent. to that provided by law. Such, gentlemen, is the plan and such the details of executing the census of the United States as now provided by statute, and I hope you will agree with me in according your meed of approbation to the practical and enlightened views of the Secretary of the Interior in his last Annual Report, wherein he expresses, in statesman-like language, the importance of this great work, and the unspeakable gain to two or more enumerations reciprocally when so conducted as to facilitate comparisons between them—a course altogether in consonance with his liberal and patriotic views on other subjects.

Having limited my paper to the subject of national statistics, I have necessarily omitted all mention of the many interesting facts illustrative of the indebtedness of mankind to the efforts of different societies, and many individuals, who have thrown great light on statistical science and done much to elevate the pursuit.

The records of the different churches in Europe have furnished some of the most important data for a long period of years, while some of this country have contributed valuable information, and many associations in Europe and America have established data of great importance. The history of these movements would furnish an interesting theme, while the moral and political effect of national statistical efforts would prove no less entertaining, and either would present points of much more general interest than the dry one which I have so imperfectly dwelt upon.

I have thus feebly attempted to present an outline of the progress of statistics, and you will not fail to have observed how in different ages, those efforts, the results whereof form the most important historical facts of the time, have originated with associations of individuals or societies like yours, and to such influences may clearly be traced the greatest public measures for improving the moral state and social happiness of man.

www.ingramcontent.com/pod-product-compliance
Lightning Source LLC
LaVergne TN
LVHW011133110826
845150LV00008B/2312
9781418194871